...and I Hate Math

Who Needs It?

Jimmy Huston

Mathematics is a game played according to certain simple rules with meaningless marks on pieces of paper.

David Hilbert

Cosworth Publishing
21545 Yucatan Avenue
Woodland Hills CA 91364
www.cosworthpublishing.com

For information regarding permission,
please send an email to office@cosworthpublishing.com.

Dedicated to Pythagoras

(and poor ol' Hippasus)

Table of Contents

Chapter What?

Why does a chapter need a number?

Why does the number 1 come first anyway?

What about zero? Isn't it before 1?

That should make the number 1 second.

See how silly math is?

Yes, I hate math.

It makes no sense.

One, two, three, four, five, and so on. Who cares?

It's numbers. Numbers are stupid.

I don't need numbers.

I don't like small numbers.

I don't like big numbers.

Chapter Next

I hate addition.

Don't tell me about apples and oranges. I don't care.

If I have 2 apples and you give me 3 more, I don't need to know how many apples I have.

Maybe I have a lot. Maybe I don't have enough. I'll be fine. It's just apples.

If I don't have any, I'll eat an orange.

I don't need addition.

I hate subtraction

If I have 4 apples and eat 1 apple, I don't need subtraction. I still have plenty of apples.

They call it finding the difference, but they're all apples. How different can they be?

(I already ate the orange.)

I don't need subtraction.

I hate multiplication.

It's just a harder way to do addition (which I already hate).

In multiplication, X means times. That's dumb. Shouldn't times be T -- not X?

This times that. Something else times anything. Who cares?

I don't need multiplication.

I hate division.

Everything always gets smaller.

Divided by 2 just means cut it in half.

Divided by 3 just means cut it in thirds. That's a little harder, but why bother?

Divided by 4 means cut it in half twice.

Divided by anything else just isn't necessary. Who cares?

I don't need division.

I really hate long division.

Too complicated.

Remainders? Yuck.

Throw them away.

Who needs long division?

I don't need math.

Chapter After That

Everybody tells me math is important.

They say that math will come in handy.

It better start soon.

Okay, some numbers can be useful, but I don't need math.

For instance....

Number 1 is pee.

Number 2 is poop.

What is number 5?

See? I don't need math.

If zero equals nothing, how does it exist?

This page has a zero on it.

The opposite page has nothing. Which is greater?

You got here without page numbers didn't you?

What page is this?

It doesn't matter.

It would've been page 20.

Maybe.

Maybe not.

I don't need numbers to know to turn the page.

Chapter Something

I hate fractions.

Fractions are just another kind of division. They're worse than long division.

Pick a number. Any number. Write it down.

Pick another number. Any number. Write it under the first number.

Draw a line between them.

That's a fraction. Big deal.

Now do it again. You'll get another fraction.

Okay, now try to add them together.

Crazy!

Stick with the big numbers.

Math is ridiculous.

Who needs it?

I hate decimals.

They're just like numbers, but they're not.

Really?

Yeah, they're just fractions without the little line through them.

But they're always tenths. Or hundredths. Or thousandths. And on and on.

Sometimes there's no end to it.

On a whole number, the zeros after the dot go on forever.

Really. Forever!

I don't have that much paper.

So they "round it off." Forget all the zeros.

I don't need decimals.

And I hate rounding off.

I spent years trying to get the right answer. Now they're telling me to just pick an easier number instead of getting it right.

Just be close. That's good enough.

I knew that years ago. Tell my teachers.

I don't need math.

I also hate word problems. They make math even worse.

If a ten car train leaves Pittsburg at 4 o'clock at 20 miles an hour, six minutes before a goose leaves Cincinnati (in Central Standard Time), how much corn would the goose have to eat to meet the train before it starts to rain?

I don't need math disguised as words.

Chapter Something Else

Math is inconsistent.

And it's full of mistakes.

Galileo made mistakes. Euclid made mistakes. Einstein made mistakes.

Add something up, you get an answer.

Add it again, you can get a different answer.

They say that one of them is wrong. Why?

Things change. Both could be right in different dimensions.

Did I mention that I hate math?.

Even NASA makes mistakes. What chance do I have?

The one and a half billion dollar Hubble Telescope was the product of the nation's best scientists and mathemeticians. After it was sent into space, it didn't work because of a small measurement error of one tiny millimeter. That's less than 1/50th of the thickness of a human hair. (The scientists and mathematicians are measuring with human hair?)

Did they get an "F" for this mistake?

No, they got over 86 million dollars to try again.

That doesn't work on my math tests.

Chapter Following

I hate algebra.

They add letters to the numbers.

And then they wonder what the letters mean. What is X?

Use better letters. Try a vowel for a change.

Unknowns? Variables? Hooey.

I don't need algebra.

And I don't need trigonometry or geometry or statistics or calculus or topology.

I hate PI.

Pi is a number that is used in some formulas.

It is the ratio of the circumference of a circle to its diameter.

And, it goes on forever. It actually never ends.

That means that you can never use the whole figure to make a computation.

So -- anything created using pi is wrong. It may be close, but it's not accurate.

So why is that okay? I've been close lots of times, but they were all marked wrong!

I don't need pi.

I hate percentages.

They're just more decimals with a silly % mark instead of a period.

Teachers use percentages to grade tests, and then they ignore these accurate scores and replace them with more vague and imprecise letter grades.

Nobody needs percentages.

Chapter How Many?

They say math is going to help me later...

Math will help me keep up with scores in ball games.

I'm just going to look at the scoreboard.

Math will help me balance my checkbook (whatever that is).

My cell phone is always with me. It has a calculator.

Math will help me measure things.

Go back and read the Hubble Telescope story.

Math will help me convert gallons to liters, pounds to kilograms, miles to kilometers, Fahrenheit to Celsius.

*I don't need to convert **anything**. I like things the way they are.*

Math will help me know how old I am.

I have an ID for that.

Math will help me tell time.

Cell phone.

Math will help me schedule things on the calendar.

Cell phone.

Someday math will help me with my taxes.

All the math I need is my accountant's phone number.

Chapter This One

I hate square roots.

In all of math, there's nothing worse (until you get to algebra).

There is a formula to get a square root, but it's a bad one. It has way too many steps and it takes too long -- if you can even remember it.

I found a better way.

To find the square root of an assigned number, pick any number and multiply it by itself.

If the answer is lower than the assigned number, pick a bigger number and multiply it by itself..

If it's higher, pick a smaller number and multiply it by itself instead.

Then try again.

And repeat.

Eventually, you'll be close enough.

You're never going to need a square root anyway.

Why bother?

I don't need math.

And they say that with square roots you can create imaginary numbers.

Yes, mathematicians say that imaginary numbers actually exist.

Then they're not imaginary.

Or, if they're imaginary, they don't exist.

Get serious.

Numbers are on a continuum, in a certain order.

Start counting.

You'll never get to an imaginary number.

You can look in between the numbers at the fractions and the decimals.

They may be small and they may be silly but they're real -- not imaginary.

You can count backwards into negativeland. You won't see an imaginary number.

But you can *imagine* one?

Huh?

That won't work on my next math test.

Math hates me.

Did I mention irrational numbers?

Why would I?

That would be crazy.

Legend has it that irrationdal numbers were "discovered" by the ancient Greek Hippasus, and that made Pythagoras mad. He had Hippasus thrown overboard and he drowned at sea.

I don't need irrational numbers or irrational mathematicians.

They say there are negative numbers.

No, there aren't.

If I have 5 apples and you take away 7 apples, do I have negative 2 apples?

Obviously, I must have had 7 apples, not 5.

Chapter Next to Last

The opposite of negative numbers would be huge numbers.

I can believe in huge numbers. I just don't care.

I don't care if it's a jillion, zillion, squillion, gazillion, kazillion, bajillion, or bazillion. There used to be a British number called milliard, but it was eliminated in 1974. Huh? Where did it go?

One of the really big numbers is ten duotrigintillion (or ten thousand sexdecillion).

They call it googol. And that's not a joke. (There are no jokes in math. No fun at all.)

The number googol looks like this:

$$10,000,000,000,000,000,000,$$
$$000,000,000,000,000,000,000,$$
$$000,000,000,000,000,000,000,$$
$$000,000,000,000,000,000,000,$$
$$000,000,000,000,000,000.$$

An even bigger number is googolplex. That's a 1 with a googol of zeros (ten duotrigintillion zeros).

They say that googolplex is bigger than the number of atoms in the universe, but who's counting?

There are math freaks who make up their own numbers and each thinks his number is the biggest. They have contests!

Skewes's number.

Graham's number.

Rayo's number. His number is so big no one will ever know how many zeros it has. Apparently it's growing faster than it can be computed.

And they won't even talk about infinity.

Mathematicians are loons.

I don't need their stinking math.

Chapter Last

I'm really glad I don't need math.

Except for money.

I don't hate money.

That's how the world keeps score.

Dollars are measured in numbers, and I don't want any mistakes!

I want to be a googollionaire.

And that's why I might need math after all.

Rats!

THE END

About the Author

He wrote a nice bio, but he put it in the wrong place because there are no page numbers.

Other Books (that you will hate) by Jimmy Huston

The I Hate to Read Book

Nate-Nate the Christmas Snake

The Dyslexic Handbook
Genius Edition

Cussing for Kids!
Etiquette for the Profane

The Attention Deficit Disorder Hyperactive Cookbook
Puzzle Edition

The OCD Funbook
Really?

The Bedtime Book of Bad Dreams
Dozing Dangerously

Baby's First Instruction Manual
How To Be the Center of the Universe

Rat BLEEP and Alien Poop
Not for Parents at All

The Big Beautiful Book of Burping, Belching, and Barfing

The Book Book
Inside the Inside Story

Why Can't Mommy Spend More Time with Me?

How to Write This Book
You're Going To Be the Author

The Amazing, Stupendous, Extraordinary, and
Somewhat Unusual SPINNING BOOK

That Damn Little Angel

The Snake Test
True? False? Maybe?

Is This Your First Funeral?
A Child's Primer

Don't Go to College, Go to Europe for Less

Dead Is the New Sick
An Insider's Guide to Senility, Paranoia, and Curmudgery

www.byjimmyhuston.com
www.cosworthpublishing.com

www.ingramcontent.com/pod-product-compliance
Lightning Source LLC
Chambersburg PA
CBHW081010120626
46546CB00010B/3096